Table of Contents

Laser Induced Breakdown Spectroscopy (LIBS)

Laser Induced Breakdown Spectroscopy (LIBS) is a very simple process to learn and grasp and is used widely throughout the science community. LIBS is a process that can be used to determine the chemical composition of some metal sample. The set up includes a few key pieces of equipment to function correctly: a high powered laser, a spectrometer, a computer, and a power source. Among each of these, more specific materials are needed; these will be covered in later chapters.

LIBS works by firing a high powered laser onto a metal sample. The laser emits so much heat in a concentrated area that when it hits the metal sample a small amount of the sample is turned into plasma. This plasma will emit different colors of light depending on the composition of the sample. If a spectrometer is used to measure the exact wavelengths of emitted light, then one can compare these wavelengths to standard wavelengths to calculate the chemical composition of the tested sample.

The basic setup is very easy, however as the small details become more and more important individual tasks must be taken to ensure optimum results. Figure 1 shows a basic setup for LIBS.

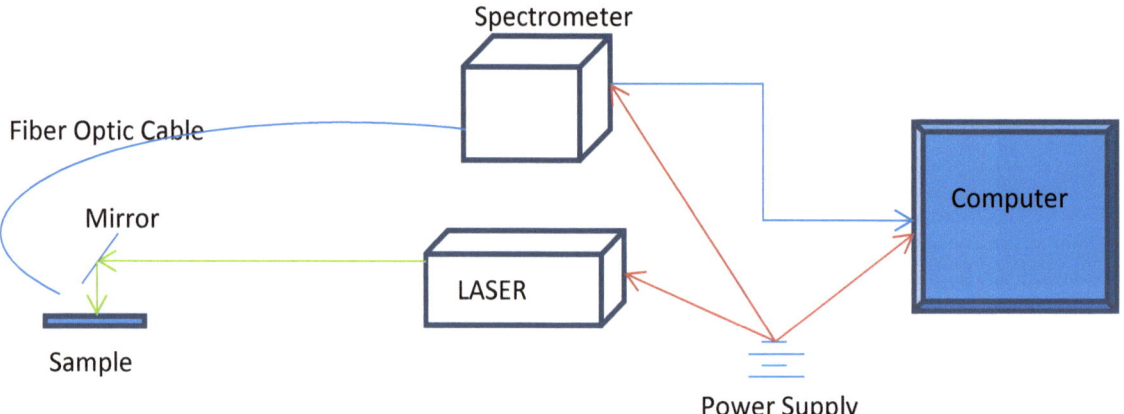

Figure 1

The laser is fired at a mirror which deflects the beam downwards towards the sample. A fiber optic cable is connected to the spectrometer and placed pointing at where the blast will take place. The spectrometer is connected to a computer which will interpret the data and graph the results.

Quick Set-Up

In order to have LIBS work at the optimum level please refer to the other sections of this manual to go through the complete set-up. Quick set-up is used if a sample needs to be taken with very little time to commit to getting an optimum sample.

Initial Set-Up

1. Make sure all necessary safety measures are taken before attempting to start working with the LIBS set-up. Long hair and loose clothing MUST be pulled or tied back. Locate the laser safety glasses and have them ready to use when the time comes.
2. Turn on the LIBS computer. Turn on the Avantes Spectrometer, the switch is located on the back of the spectrometer.
3. Start "AvaSoft 8" on the computer. On the loading screen it should show that 4 spectrometers are available and activated, if not, shut down the program and turn off the spectrometer, then restart the spectrometer and program.
4. Once the software is running, go to the LIBS sample area and open the door. Place the sample to be tested directly on top of the rotating device.

Alignment

5. Use the small red laser pointer to simulate the fired laser by placing the laser pointer at the aperture of the high powered laser. Remember that you must be extremely careful when using this simulated laser. You must be sure that the laser is completely level and straight. Any deviation in these two parameters could result in an incorrect alignment.
6. If the mirrors are at the desired positions for alignment then proceed to "Camera Set-Up".
7. If the mirror and lense are out of alignment then adjust them to the point where your simulated laser hits the desired position. REMINDER: Do NOT move the lense up or down as this will bring the laser out of focus.
8. To achieve the preset focus for the current height of the laser the top of the sample should be approximately 8.5cm from the baseplate. If this needs to be adjusted, use the screw jack to change the height of the sample.

Camera Set-Up

9. Turn the camera and monitor on.
10. Make sure that the coaxial cable between the camera and the monitor is connected and functional. The cable should plug into the port farthest on the left for the monitor and farthest on the right for the camera.
11. Switch the monitor to Video A.
12. Move the camera up and down until the bottom left corner of the monitor shows the sample in focus, this should be approximately 7.6cm if the camera lense is set to .7 by rotating the lense all the way clockwise. NOTE: The entire screen will not show a focused sample because the camera is at an angle.

Fiber-Optic Cable Set up

13. The fiber optic cable is attached to a rotating, pivoting device that you can use to align the cable to the correct position. In order to do this, the laser must be fired so the exact position of the test can be determined. Follow steps A through J to fire the laser for the first time.

Firing the Laser for the First Time

   a. Make sure the person who will be operating the laser has the laser safety glasses on.
   b. Turn the key on the laser to the on position.
   c. Turn Stand By on. There will be a beeping noise that lasts for approximately 10 seconds. This means that the laser is warming up and should NOT be fired at this time.
   d. After the beeping has gone away you are ready to fire the laser. Turn the frequency to 1Hz.
   e. Turn the shutter on.
   f. Wait 5 seconds.
   g. Turn on emission. After doing this the laser will fire. Since this is only for alignment purposes the laser only needs to be fired once. After it has fired turn the emission off.
   h. Close the shutter.
   i. Turn Stand By off.
   j. Turn the Key to the off position.

14. Proceed to the sample area and observe where the sample was fired upon.
15. Adjust the fiber-optic cable stand so the tip of the cable is pointed directly at the mark that the laser just created. The distance between the cable and the mark should be approximately 1cm.

Realignment

16. At this point it may be necessary to adjust the lense and/or mirror again to bring the laser into focus or to have the laser mark show on the monitor.
17. By rotating the mirror and lense in the x and y directions this should be able to be achieved. NOTE: Tend to stay away from adjusting the camera unless it is absolutely critical to be able to see the mark.
18. Once the mark can be seen easily on the screen, observe the shape of the mark. If the laser is being focused properly the mark will be a circle. The more unfocused the laser is, the more ellipse like the mark will be.
19. If needed, use the screw jack to adjust the height of the sample and re-fire the laser until the maximum focus is achieved. NOTE: If the sample height is changed too much you may be required to readjust the camera.

Taking the Sample

20. Turn the rotator on and using the monitor, be sure that the sample will consistently stay in the path of the laser.
21. Close the door to the laser and proceed to the computer.
22. In "Avantes 8" start the sample process by pressing "start".

Firing the Laser for Taking a Sample

    a. Make sure the person who will be operating the laser has the laser safety glasses on.

    b. Turn the key on the laser to the on position.

    c. Turn Stand By on. There will be a beeping noise that lasts for approximately 10 seconds. This means that the laser is warming up and should NOT be fired at this time.

    d. After the beeping has gone away you are ready to fire the laser. Turn the frequency to 5Hz.

    e. Turn the shutter on.

    f. Wait 5 seconds.

    g. Turn on emission. After doing this the laser will fire. At this time you may proceed to the computer to watch the samples being taken. Every time the laser is fired the spectrometer will take one sample. At any time in the process you can click stop to stay on a current sample graph.

    h. After you have ran the test for the amount of time wanted, turn emission off.

    i. Close the shutter.

    j. Turn Stand By off.

    k. Turn the Key to the off position.

23. Using the appendix in the Handbook for Laser Induced Breakdown Spectroscopy and the graph of the sample, determine what elements are in the sample by comparing the peak of the graph to the standard wavelengths for each element.

The Laser, The Mirror, and The Lense

These three key components can be very tricky and frustrating to work with at times. Luckily, the laser is always in a fixed position but the lense and mirror can be bumped or moved very easily. When working with components inside the enclosure be extremely cautious of these two components as you only want to move them when aligning for a test.

The mirror is the easier of the two to deal with as it only changes the direction of the beam. However, the lense can change both the direction and focus of the beam. Figure 2 shows a general setup that was found to focus the beam on the sample almost perfectly.

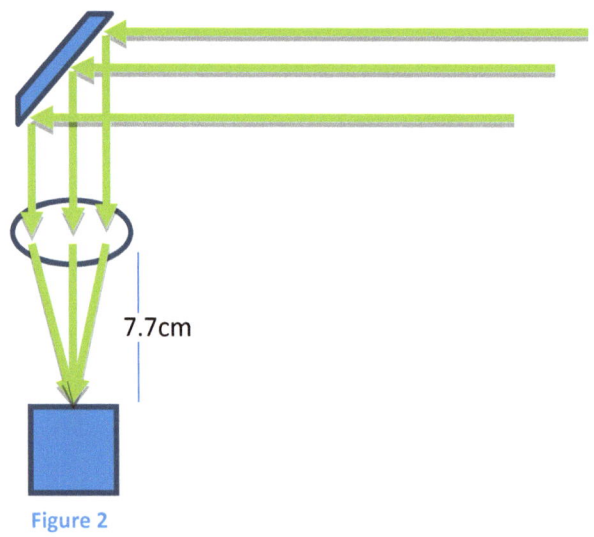

7.7cm

Figure 2

The Spectrometer and Fiber-optic Cable

This seems to be the easiest component to use and set up. Once you fire the laser once for alignment it is very easy to adjust the fiber optic cable to the desired position. If for any reason the rotation base for the fiber-optic cable stand needs to be removed then first detach the stand for the rotator by loosening the latch the unscrewing FROM THE BASE as to not twist the fiber-optic cable. Then unscrew the base and place in the new desired position. Re-screw everything into the correct position.

The Video Camera and Monitor

One method that was found to work rather well to focus is the laser is to use the focal point of the camera lense. This can be very useful instead of having to readjust lenses. Of course the first time you take a sample in a series of tests the lenses are going to have to be adjusted and aligned to the sample. When a different sample is used, unless the height is exactly the same as the last sample, the laser is going to be out of focus when it reaches the sample. There are two ways to solve this problem. You could readjust the lense and do test shots to determine where the focal point is, or you could use the cameras focal point.

The camera lense focal point is the same as long as the focus is left the same. This means that if the first sample has been put into focus with the camera and the focal point has been set with the lense that means, whenever a new sample is tested, all you have to do to focus the laser is raise or lower the height of the screw jack to bring the face of the sample into focus on the monitor.

Sample Placement

This is one of the most important pieces of the LIBS process. When continuous tests are run, the sample MUST rotate. This is due to the phenomenon that if the laser fires upon the same spot twice then the spectrum produced will be very weak. To solve this issue a simple rotator is used to move the sample very slightly so that the laser fires on a fresh spot every time. When doing longer tests it may be required to turn the laser off and move the sample slightly to start a new circle of marks.

Avasoft 8.1

Avasoft is a rather simple software to learn and very user friendly. Luckily, when running LIBS tests there is very little you have to do in the program. When starting the program be sure to turn on the spectrometer BEFORE you try to open the program. If you fail to do this, the program will not detect and spectrometer and start causing problems. Always remember to start the spectrometer by clicking the start/stop button before you begin testing. The Start/Stop button is shown in Figure 3.

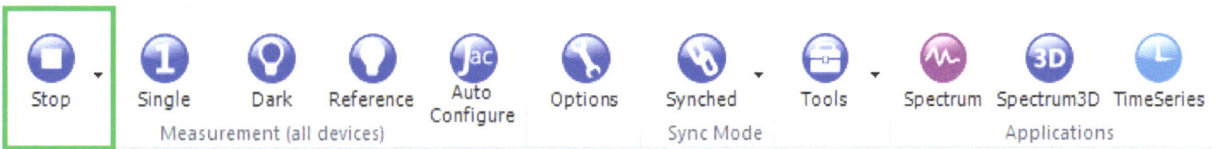

Figure 3

There are a few options that deal with testing that can be changed to meet the needs of a specific test. The standard settings for these options will work rather nicely with the test, however if you would to change some of the settings to get a different output the options are readily available.

External Trigger Settings

The External trigger is placed inside the walls of the LIBS set-up. The purpose of this trigger is to tell the spectrometer to take a sample. When the laser fires, the light that flashes from the creation of the plasma goes into the optical sensor located on the trigger and tells the spectrometer to take a sample. The External Trigger Settings are located under the options button as shown in Figure 4.

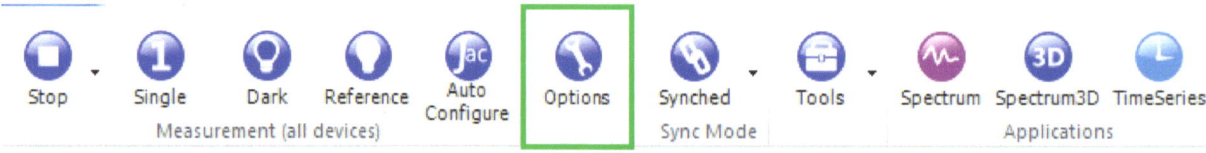

Once the options tab opens you will have a couple of options to choose from. You will want to click on External Trigger Settings as shown in Figure 5.

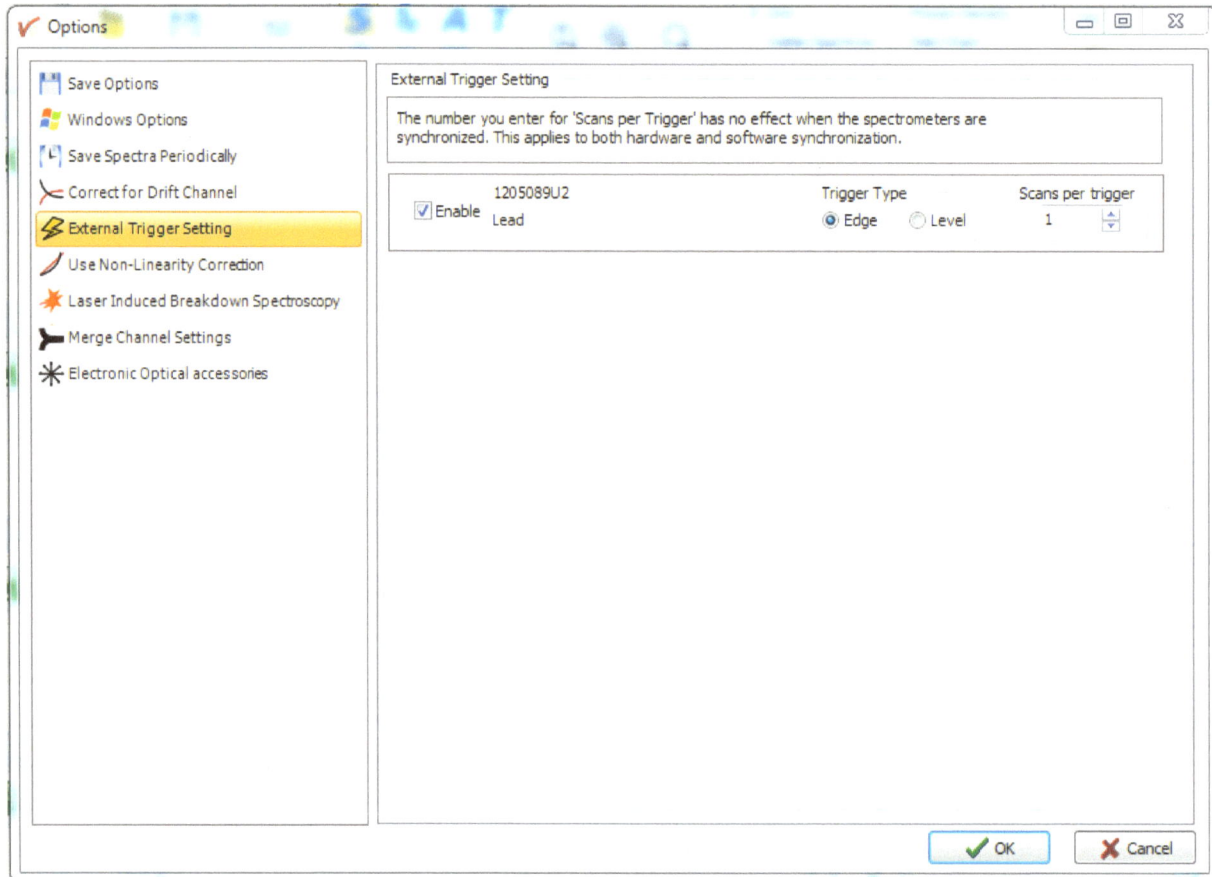

As you can see from Figure 5, you have the option to enable or disable the External Trigger. When running LIBS test you will want the External Trigger enabled, for without it, the spectrometer will take continuous samples instead of just when the laser fires. You also have the option to change the number of scans that are taken per trigger. As a standard, 1 scan per trigger is used.

Saving Dark and Reference

Saving Dark and Reference data is an optional step, however, it does have the possibility of creating better data. To save the dark and reference data for all the spectrometers use the Dark and Reference buttons on the toolbar as shown in Figure 6.

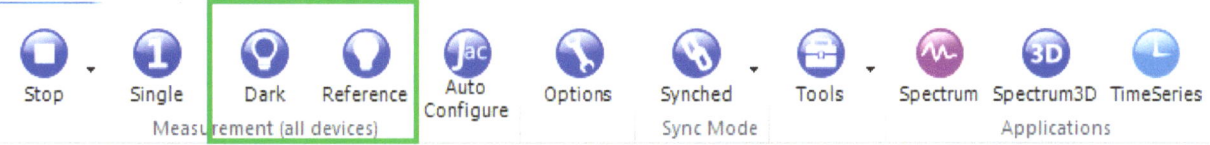

Figure 6

To save the Dark data, make sure the door to the LIBS is closed and the laser is OFF. Click the dark button and the spectrometer will save this dark data. If the data is saved correctly then each individual dark data button for the spectrometers will turn green as shown in Figure 8.

Figure 8

Figure 7

To save the reference data, turn on the laser then click the reference button. Once again, if the spectrometer has saved the data properly the each individual reference button for the spectrometers will turn green as shown in Figure 7.

Integration Time

Integration time is the time between when the trigger is activated and the sample is taken. The default integration time is 5ms and can be changed for sample preferences. Each spectrometer must have their integration time changed separately, to change the time click on the desired spectrometers current integration time as shown in Figure 9.

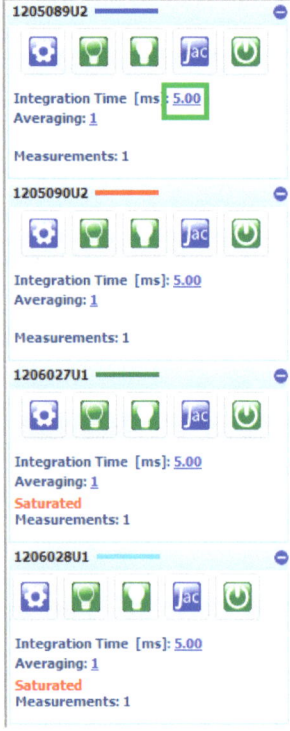

Figure 9

Exporting the File to Excel

After you have obtained a spectrum you would like to use to determine what elements are present in your sample you can export it to excel for further evaluation. To do this, click on the files tab as shown in Figure 10.

Figure 10

Then click the export to excel button as shown in Figure 11.

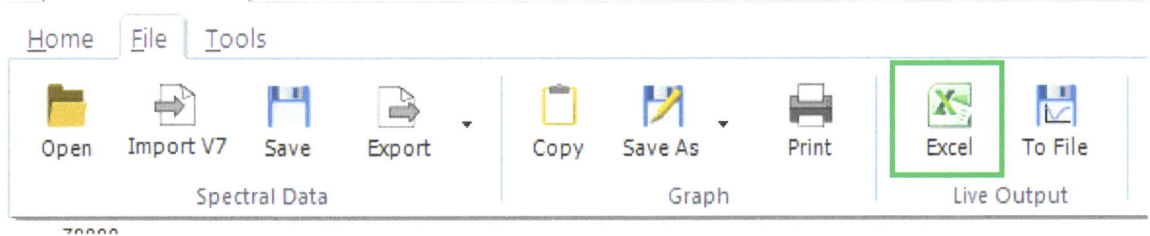

Figure 11

When you click this button, AvaSoft will convert the data from the spectrometer into an excel data file. This file will open in a separate window. The next section will cover how to create the graph to evaluate this data.

Interpretating the Sample

Open SampleTemplate.xslx located on the desktop. From the exported excel document copy each of the spectrometer data into the Sample Template. Each of the spectrometer data will have been placed on different sheets of the exported excel file. Remember when you paste this information into the Sample Template to place it directly where the current information is, as the graph is already set to use certain cells for its data.

Once you have finished copying the data into the Sample Template you will see that the graph has changed to show your data as shown in Figure 12. Save the File as your own by clicking Save As. DO NOT click Save, as it will overwrite the Template file.·

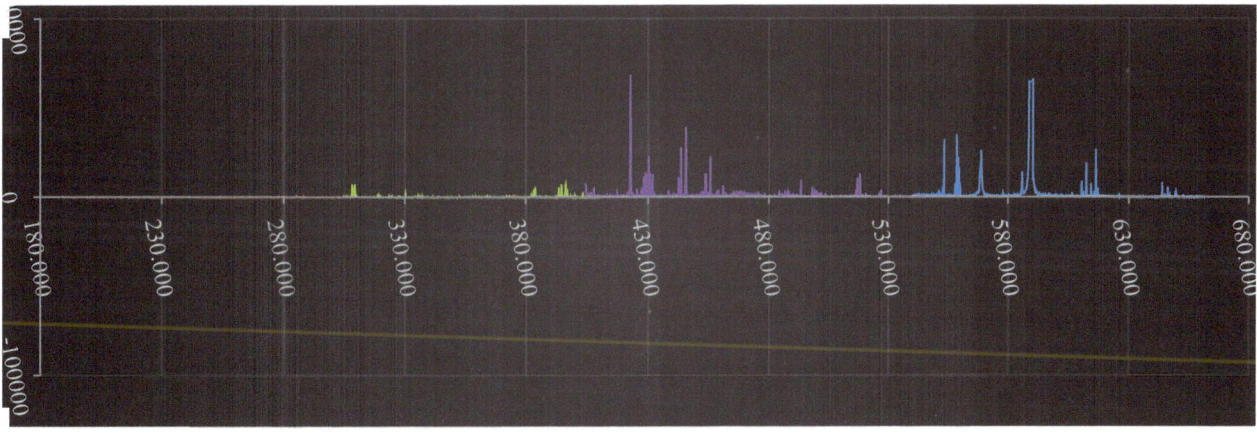

Figure 12

From this point you can place data labels at each of the high points to make it easier to determine the exact wavelength for which the peak occurred. Now using the Handbook of Laser-Induced Breakdown Spectroscopy located in the lab, you can determine which elements are present in the sample you have just tested by using the Uniform Detection Limits. These limits are located on page 263-265 of the LIBS handbook.

Notes